The LARGER EARTH

Descending Notes of a Grounded Astronaut

Permeable Press • 1996

The LARGER EARTH

Descending Notes of a Grounded Astronaut

A Cycle of Poems
David Memmott

Collage by
Freddie Baer

Acknowledgments
Various poems in this collection have been or will be published in some
form by the following publications: *The Magazine of Speculative Poetry*, *NRG*,
*Star*Line*, *Xizquil*, and *Co-Lingua: 15+ Years of NRG*

A Permeable Press Poetry Book

First Edition

ISBN: 1-882633-18-0

Permeable Press
47 Noe Street °4
San Francisco CA 94114-1017

E-mail: bcclark@igc.apc.org
Permeable Press on-line: http://www.armory.com/~jay/permeable.html

Write for a free catalog of our other fine publications.

Also available from
Wordcraft of Oregon
P.O. Box 3235
La Grande OR 97850
wordcraft@oregontrail.net

Cover illustration by Freddie Baer
Cover design by Brian Clark

Rather than living exclusively in disembodied heavens or "separate realities," we can broaden our horizons in this world, conceivably, and realize new kinds of life in contact with more and more dimensions of the universe. Our further development, in short, would open the world we now perceive rather than disengage us from it.

- Michael Murphy
THE FUTURE OF THE BODY

...for every man shares the responsibility and the guilt of the society to which he belongs.

- Henrik Ibsen
Introduction to PEER GYNT

Sure, we use science, we use our knowledge, but in doing so we transform ourselves.

- Brian Goodwin

For this work, you should employ venerable Nature, because from her and through her and in her is our art born and in naught else...

- Rosarium philosophorum

CONTENTS

WHILE PLAYING ON THE SHORES OF THE HOLOFLUX THE GROUNDED ASTRONAUT POPS A WAVE FUNCTION AND BLISSFULLY BRINGS A POEM INTO BEING

Above us the giant
chews on his jargon, (a curious
cud of conditions);
a jumble of technicalities
drools
in black and white like spittle
down his double chin.
Beside this giant
we settle for mediocrity
not thinking how each contest
we concede marks our gradual decline.
A meteoric fallout of half-digested slogans
like shards of looking glass strewn
on thin ice for jaded pawns
reflect a skewed reason,
splinters of blue
cracked from heaven exploding
somewhere beyond fractured mountains,
crumbling
before the purge of perfection.

The wizened old crones on the beach
have their own reasons for swabbing bone-deep wounds
with rock salt, pointlessly digging holes
in the sand which the waves, pointlessly repeating,
only fill back in again.

Eccentric old women
crumble
before the purge of perfection
and scratch their bald heads,

wondering
        why such small progress
        after a lifetime on their knees,

wondering
        over splinters of blue heaven until
        the twilight time
when you see them on their screened porches
rocking in their rockers

  polishing the relics of dawn.

In their endless effort to renew
      the world,
through their Herculean task of
      simple recall
marking their own meteoric and irreversible
      decline,
I see myself beginning.

I see myself beginning by the quick stream,
watching the current ride
over rocks.
Soon the swirling eddies and backwaters
carry me, a spinning leaf painted with fall,

plunging

with gravity toward
the eventual ocean,

emptying into all that is,
  always without finality,
  always beginning again,
a round,
    a loop,
        an ecstatic cadenza.

Open us now, siblings of the five worlds,
five pregnant women ready to give birth
waiting at the open gate,
pregnant with possible worlds where
a crack of sky or a piece of shell
is a cavernous conch through which dreams enter

   moving us

like a song
slightly offcenter,
breaking
the stasis of cold
unremembering.

For me, the obvious is no longer obvious,
but a cud of conditions, a jumble of jargon.
The rock
is no longer solid rock
but space   within   space,
opening at dawn and twilight
inner world wombs
caught by the
quick stream.

The tools of creation are composed of
       light,

       light of thin ice reflected in looking glass,
       bred in dank and dark places under
       rocks and stairwells,
       in the glistening glidepaths of slugs,
       the mystic meanders of luminous wormtrails.

Between dreaming and waking,
by the quick stream where current rides over rocks,

in the visitation of many forms:
> spirals,
> cones,
> crystals,
> branches,
> fractals,
> Mandelbrot sets.

There are holes in space both near and far.
The elasticity of reality conforms
   to idea,
to the euphoric locus
> of a sedimentary mind
where time and space s t r e t c h and curve
until today touches tomorrow and knowing one
you know the other.

Each star blossoming
in the night sky implies a swan
born of collective thought
as each thought by the quick stream
wings its way
to the juncture of mattereality
bobbing in the holoflux beyond the siren wail
of old women
crumbling
before the purge of perfection.

THE GROUNDED ASTRONAUT HIDES
IN THE MAZE OF THE PENTAGON
TO AVOID RETURNING TO SPACE

I am
small,
so very small,

adrift

like a message in a bottle,
like driftwood with facial hair
weathered away,
my voice
f/r/a/c/t/u/r/e/d
as a stone tablet buried in the ruins
of a b o m b e d o u t cathedral.
There was a time, perhaps a dreamtime,
when I shouldered my way
through the crowd
to stand elevated
above menial men,
a swan in bloom, composed of light,
yet from that height
I coldly stiffened into a blindly
minor god
longing only to diminish   again,
perish and pale
to no mean
significance.

Grant me now anonymity,

erase my name from the honor roll,
from all the gold-plated plaques

and social registers.

I wait below for some flyboy   to crash
so I can pull him from the wreckage
and welcome him home.
We gather at the site and pass the word
silently among us.
Do not overachieve,
do not fly too high,
stay low, go in under the radar,
be inconspicuous,
do not attract too much attention,
there are soulcatchers out there
with an eye for bright boys and free-fliers.
Sharks patrol these waters.
They will sniff you out.
Choose to be no one.
Pass the trophies to someone else;
let them become the target of adulation,
idolized by a cult following
who come to curse you because they insist
on your perfection.
Tell those who parade
through bannered avenues   sitting
on the backseats of white convertibles
under a ticker-tape shower of half-digested dreams
that you retired your   will and rectitude
to sit on a sunwarm rock
by a quick stream.

THE GROUNDED ASTRONAUT
ON THE EDGE OF A BREAKDOWN
LOBBIES CONGRESS TO REGULATE
THE EXPLOITATION OF FOREIGN BODIES

The breakdowns of spacesick sailors and blacked-out pilots
presume a humanity not tolerated in these impermeable reaches.
              Between the cat   and cathouse
                      fly the fast and furious.
              Charioteers of progress   chewing
                  on cheap cheroots while the sexless
              dowager   pricking the fingers of donors
                          sucks their blood
                              into tubes,
                        sucks their blood
                        and swoons at the sight
                of a shirtless terraformer with red
                mud in the tread of his boots.
In the blood flows the lost oceans of Mars.
In our bones, a marrow of stardust.
We send these missionaries to far flung reaches of space
to assimilate all differences, erase all history but our own,
transform all otherness into gleaming facets of Self.

              Forgive me now, esteemed brothers,
              I was simply curious
                      in joining the parade,
                  how can we fly so fast and furious,
                      chewing on cheap cigars?
We make the universe our frontier and look to the stars
  glimmering in the distant reach of our projections.

Between ambivalent latitudes of promissory politicians
           so spasmodically masculine in their dodge,
           how is it we can stand

                 counterpoised

between the sexless dowager sucking
            blood into boots

and the slick and stainless broker
        sinking his teeth into   soft red earth
        on consent of forms in triplicate?

Despite mountains of consumer warnings
        of corporate madness
                swilled into the brew,
        the copious crew protects its appetite.
Integrity
                is catabolized by compromise.
Madness sinks its teeth
        into our sleep
o p e n i n g
a wide abyss   between
day   and   night
        and in our sleep the hunger
murders ambition
leaving avoidance   as the planned
poison of partnerships
        and last year's integrity
served up on platters by sore-headed heathens
counting converts in their sleep,
        counting forms in triplicate
and   warnings never heard.

Alien thoughts recur, small novas
bursting in the darkscape of mindtime
and our breath ingests the residue
of Atlantis,
        expelling a new age.

Afterhours
        joyriding with some dictator
                counting converts in our sleep
forebodes a dead finish
        for some baldheaded aide
                living a lie,
                peddling his wig for a Ph.D.
                and a recommendation,
                exaggerating the size of his penis.
The broker is slick and stainless
        with a secretary sworn to secrecy
                sucking blood into compromise,

counterpoised

between lost tribes   and last tribulations
        where madness
                murders ambition.

                Yet, the trembling gestalt of who we are
        is constantly cannibalized by endogenous renegades
                feeding on a vast, subjective doubt which eats through
        the onion skin of identity leaving us no center.

Cowardice
        is a form of jaundice that bends
                the rules to serve a few.

A nocturnal network of hooligans shrugging their shoulders
at the sound of machines   breaking down   hang around
until the springs are sprung   with yellow skin warping
        in the sun
faking fibrillations that throw
                the world
                into a fit;
they winnow the past through raw rubs
        of s t r e t c h e d truth to fabricate

a history of hate
out of monkey charades.
All a part yet forever apart, irreducibly whole
yet fragmented by the very power which drives us
to the constellated limits of our space/time.
A form of jaundice sucks blood
from Mars, terraforming the red earth,
a nocturnal network of hooligans
projected in holograms
chewing on their cud,
flying fast and furious,
tearing up runways and reeling
at the azimuth where spacesick sailors'
white tallywhackers melt into fog
as a star collapses
around its own gravity.

THE GROUNDED ASTRONAUT WITNESSES ANOTHER SPEED JOCKEY GO UP IN FLAMES

They nowhere go uninjured,
these thrillseekers with gasoline on their breath
and sparkplugs for eyes,
self-immolated tumblers
casting their lot with
the freaks of faith.
We didn't start the fire
but we all suck on the ashes.

All the tools we possess are
weapons.
The last warrior, too slow
for suicide, sits on a warm barstool
immersed in the ghost
gestalt of stranger bodies
toasting half-digested slogans,
(a blindly minor god)
learning the hard way the lesson
of humility as object of ridicule,
taunted for his fear of flying.
Some jokers would kindle their kin
for the mournful cadence of automatic applause,
laughing on cue
like a pack of hungry hyenas.

New masters go on breaking slaves.
We didn't start the fire

but we all suck on the ashes.
When the willows steal
the harp from the wind

the song grows sullen and still,
[captive] in the cosmic house
of stairs spiralling up to where
a body-conscious giant chews on
a cheap cheroot and sips a tall cool one
while the boys on the street, hardwired
and hardlucked, singing hard times
are hardly times at all, say no
to everything they can't mainline.

How does the heart
endure such drought as this?

All through the circus town
poor refugees from the sideshow
slip into niches to snatch a fitful sleep,
crumbling
before the purge of perfection.
The mind-altered acrobats – poised
on the edge of an ill-fated leap –
spell a fortune for the prophets of profit
somewhere down the line,
somewhere beyond the f/r/a/c/t/u/r/e/d MOUNTAINS
where there is some sense beyond the senses.

How does the heart
endure such doubt as this?

Surrounded by heavy heads of wheat so ripe
a spark from a spinning wheel
could set the world ablaze,
they nowhere go in moderation
but wildly thump the thin metal of their dreams,
stripping their gears.

Young champions, lacking the grace to grip
the essence of life on earth,
stand hypnotized and honored, counting converts in their sleep,
plotting a strategy for world dominion
that will leave thousands of refugees
living out of bags.

A body wave of well-oiled sun-worshippers
roll together in the sand, coupled like hydraulic crabs
so new masters can
go on breaking their slaves,
teaching right from wrong
by breaking a few arms.
The antics of angels
once favored
misty blue ridges over the final score,
the promise of heaven over the nuclear dream
of b o m b e d o u t cathedrals and melting MOUNTAINS
laying out boundaries for the wild in the west.
They nowhere go uninjured,
beaten down
by make or break until each crack
in the leather of their smiles is pondered
into metaphor, a sad caricature
of the conquering hero.

Sitting in grandstands, we roar our approval
as speed jockeys crash into walls, bursting
into fireflowers on the sweltering track,
stiff-legged Frankensteins dancing on fire,
falling and rolling in grass as pit crews
come running with their extinguishers.

Whether it is human error or mechanical failure,
suddenly the blackout is forever.

Do we know how w i d e
the new world is?
How much light it lets in?
Can it be measured with these tools we possess?
All our instruments are lie detectors
detecting bone-deep wounds that never heal.
I see myself
beginning
in the weary shadows of old acquaintances who
see each other for the first time in years
by breaking the spell of chemicals
and fast food,
who finally face the fact
of many worlds,
many lives,
unfolding
like the petals of a rose.

THE GROUNDED ASTRONAUT CONSIDERS A FEW FUNDAMENTALS IN HIS ROLE AS OBSERVER

In the pearl corners of Palestine,
>> the bleached milk of dull-witted
>>> mammals too slow for suicide
>>> flows in clean basins,
>>> simple and pure.
Yet out of lip service convulses
>> a few drops of ecstacy, enough
>>> to blend with the blind and soul-smarting
>>> vision seekers some promise of progress.

For me, the obvious is no longer obvious;
> the rock no longer solid rock.
Those cathedrals built without foundation lost touch
> with the earth and closed their doors on us.
We scrawl graffiti on its walls but cannot escape the street.
We wrap ourselves in bell clappers
> and mistake the soft knocks for stars in our eyes.
How is it the lucky charm for some
> is bad news poison to others?
The gentle persuasion of stars in our eyes
> lies armed today with dynamite in our teeth,
> each word detonates
> blending with the blind promise of progress.
Yesterday's dime novel edifies
the pink pardon of some temptation
> bred
> in dank and dark places.
> It presses soft against the stiff necks
> of dignitaries and I cannot blame them.

Their cool composure need not wash off
>> so easily as lovestains (and oh how love stains),

offending hands
        emerging from the tuck and roll
        of worried sleeves, sober again, in control now
    ready to resume official business
    standing apart again as impartial observers.

Down from the grave
        site their shadow swells, swilled
        with a passion for armor clinking
        in the wells of wealthy clansmen on night shift
        leaving no trace of Christian grace to obscure
        what is obstinate in their easy minds
                until the twilight time,
                until I hear America
                        screaming

                under a burning cross.

Stark, stark,
    so dark and dry,
            the dung heap hepped with ideals,
            the last remains of blundering bluecollars
their sweat-drowned lungs
    aqualunging
            from a grey archipelago of steaming slag
                    by the rundown rivertown,
                    by the once quick stream
        where their children play.

All night
        the rites of glass blow a hollow wind
                out to see how surly waves survive

                with hairy forearms pushing back the shore,
                        offending hands on wasted cornucopias
                        collapsing with gravity
            beyond the melting mountains in the nuclear dream

              - in control now, ready to resume.

A shoulder becomes a pillow
   supporting the palls of purple faces
   of divinely average men
   with no place to go.
When they are gone, you motion
   for a bird's eye view
         of fabricated national parks and sacred places
   where their brain-damaged children can be unleashed
in pallid droves
   as if a moment of Bacchic abandon
in some programmed cyberspace can compensate
for the loss of a whole generation.

For me the obvious is no longer obvious,
       but when they're gone,
       they're gone;
           when their work is done,
           we can never go home
           for we have no home;
           it was buried
           under the promise of progress.

77 S7 SA 37
505 DU 65

THE GROUNDED ASTRONAUT SEARCHES
FOR THE GODMAN DOWN BY THE TRACKS
HAVING FALLEN OFF HIS HIGH HORSE

A ragged column of window watchers
sniff the rusty rails
to where sputtering constitutes a language,

     to where all the lost in a lost land
gather like raindrops in shallow pools of reckless
        humanity.
The godman goes nameless here among the wild,
      content with instinct and impulse,
running with the hounds of heaven,
working back alleys til dawn.

I watch them come and go, social droppings
left behind in streets of slums chewing on resentment
for drawing the lot of life. Myriad deadends
of dare doers in a show of hands,
     their fresh stigmata
bleeding onto white shag in classic limousines
while being chauffeured to a clinic to donate
internal organs to organizers.

Pouring redhot cauldrons of steel
     in factories that fill the dark beyond
       with the sharp retort of angry unions
        casting strike votes,
we worry in wait
of endruns
round the picket lines.

When they're gone, who's left behind?

Hoary, hoary, who gets the glory

if the famous mother of mercy
cannot spare a dime?

The dull dud   brushing a high hat
          in the local   high school   stageband
cannot hear   the barn door creaking
and never makes it   to the crossing in time.
He may never know what he missed,
not making it in time to see
the oxen people bearing their yokes,
reciting rhymes in unison,   working to pay more
               deductions
and living only to retire   in time.
If he had made it to the crossing, he might have been
          spared.
He might have seen me walking on the moon
  a refugee of earth
looking back as through a microscope at some microcosm
alive with movement, changing before my very eyes.

For me the obvious is no longer obvious,
We nowhere go uninjured.

So I sit beside the quick stream
          before the purge of perfection
               living out of a bag.

When winter cracks the calendar   into f/r/a/c/t/u/r/e/d days,
the waltz wars
          somehow worried me more
than the sound of ice breaking,
the thin ice of

                    mirrors reflected in mirrors,
giving back only the surface gloss of painted lips.
I wonder how long the paint can stand up to this wind?

Will this box of wet weasels I carry home
    be slick enough tomorrow?
Will the paint still be bright enough?
Will the gloss fade and crack?
My drooping gray whiskers reveal what congeals
    in my marrow
between stains of pie alamode and denture cream.
I stood above it all with no compromise in my eyes
directing traffic. Seeing them now is seeing myself
as I could've been, breaking the stasis
                    of cold unremembering.

Old Bluebeard of the dominant sun, my father,
is dead and done;   he crosses
    and recrosses   his own tracks in expanding deserts
    and ends pathetically   a neglected prophet
and pathetic patriarch   without authority
snaking sidewise,   scratching verses in the sand.
But his specter keeps haunting me.
He knows my secret; I failed
            to protect my mother.

Waxing the wings   of a flying horse
    that eats the sun,
but still too tense to escape the pull of gravity,
too bound to my own fears to climb on its back and let go
of the reins, holding onto its mane as it canters
    into the air, leaping from a sheer cliff,
falling
to certain death but then rising,

        flying,   conquering the sky.
    The sun is his center,
    his breath is light,
    his hooves stamping the black void
            with stars, galloping galaxies,
                spiralling hordes of flying horses, sense

beyond the senses,
reaching my ears.
How can I describe it to the man on the street?
What can I say to the men and women left behind,
living more and more in a world made for machines,
    where technicians with dangerous flat affect
hit on their computer like it wore a skirt?
What can I tell my neighbors grasping for the golden calf,
yet believing the golden age is past? I can say nothing
not better said by the sack of bones
    sitting in the dust
    waiting for death from starvation.

        Yet vain are the valiant. They fall
                into falsehood with all the convenience
of Goldcards and strongboxes
            and sell
poor souls   on the promise   of heaven.

One by one they devour all the verities
            and it sometimes snows hard enough in July
to spell the word
                    that means more
                            than man.

THE GROUNDED ASTRONAUT INTERRUPTS
THIS PROGRAM FOR A WORD
ABOUT YOUR SPONSOR

When the last dance strikes
a hot lambada with cold
pastrami feet
slumbering slightly to the tune
of tomorrow, the dancers see
blue as green
in the misty air
beyond melting mountains
beside the quick stream.

In that twilight time on some remote island rising primordial
from the mist, a half-dream, half-reality merged in a
subterranean chamber, a real Genesis beast, amphibious, come
out at last, squinting in the sun. One by one they accumulate,
restlessly longing, lost in a jungle, in shadows of leaves where
the singing of angels can be heard in a far off canopy; they
accumulate and their movements accelerate, clashing like the
great dischords of Stravinsky.

Three monks in disguise
watch from a lighthouse
on a cliff as waves erode
the ground from under them.
Their bodies throw
shadows on the night sea
but they cannot get out of the way,
let the light through.
Their crippled light falls on lechers
falling down drunk on crippled piers
collapsing
around their own
                              gravity.
The heavens have closed their doors to me.
I am not the man I used to be.

I lobbied for pitiless, self-made men content to suck blood for a
living, feeding rat poison in lollypops to children on the streets.
Then complained wildly of all the   [limits]
emptying my deep
pockets in favor of   an elk or a goose,
                or a swan composed of light.

These picture windows
shaded by crystal blinds
only magnify the mugshot   of a scorned
                self-perpetuating perpetrator
                        of unspoken crimes.
The disparities are parodied,
paradoxes unresolved
by efficient equations   and perfect formulas,
formulas developed   in the purge of perfection
by the bosses   of a hundred doormen
   standing to gain.   Not a simple approval
                        for the humps and dumps
                                of death mounds
sleeping like dragons in the playgrounds
where children sicken without hope of cure
   while hounds of heaven   counting forms in triplicate
offer only a jumble of jargon, a cud of conditions.
What A FRANTIC FARCE I Have Sponsored!
The toxic lives
   and plastic dreams of factory workers   wasted
on the promise
                of freedom in America.
We stumble into fog
shedding our wool.

Do we know how high the new world is?
What walls we must climb over?
Who can shed the bones of history
to wed and bed the blushing bride

in the damp earth after an acid rain?
Whose nails bleed the body of its spirit
in a crosswit battle of myths that no longer fit?

"Not I," says the sponsor, "Not I."

And what was the wise man's biggest concern
    as he sucked your blood and swooned?
That the last crude oil dumpling
    would roll off the table and split-splat
    on a dirt floor, squandered on a blindly minor god
    who flies fast and furious to dank and dark places
    between dreaming and waking by the quick stream.
At two bucks a bite
    one opportunity to salvage a soul
    costs no less in time than swallowing
    a bowling ball rumbling down the fast lane.
The mere suggestion of heartburn
    is meat enough to start another revulsion.
Who will keep score when the bombs go off?
Who will sing in tune with scales of tomorrow
as the dancers see blue as black?

"Not I," says the sponsor, "Not I."

THE GROUNDED ASTRONAUT ON A PILGRIMAGE TO MECCA LOOKS FOR NEW RITUALS AMONG THE RUINS TO HELP RECOVER A LOST GENERATION

Begin

with a blue musk ox
in hyperdrive
chasing a mirage to Mecca,

> or a faked-out farmer
> on a pilgrimage
> wearing designer jeans
> into a Moslem mosque,

or even a tattooed truckdriver
who sleeps standing up
for fear of bedbugs
infiltrating his indigo warmers
PROPERTY OF
which he flashes
at unwary women while jogging
> up a mountain trail
> for a peak experience
> in the High Sierras
> the first week of June,

>> and you will find yourself
>> filling green bottles
>> with old vinegar
>> and calling it wine.

The world makes liars of us;
> we settle for mediocrity
>> skating on the thin ice of skewed reason.
> You will find yourself calling it wine
>> when you break it out

at Christmas and pass it around
                    to the speed jockeys,
        self-immolated jugglers tossing up their brains,
a whole lost generation of body-conscious
                        blindly minor gods
drinking bleach and sucking on ashes,
                wildly thumping the drums of hell
                    believing its the door to heaven.
They nowhere go uninjured
  sacrificing their limbs
                    for that look in their eyes.

        Oh, come on, you know them.

        They are your relatives
smiling and relatively   smacking their lips
        though their stomachs revolt with a sour magma
            burning their throats.

They go ahead and drink it anyway,
        skidding on pangs of sentimentality,
            not wanting to offend,
        hoping this once they've been handed the truth.

Who's to blame anyway?

"Not I," says the minister.
"Not I," says the priest.

        With their new religion of on-line electric nightmares
                and the narcotics of plugs and connectors,
                they hit on a computer like it wore a skirt,
                another high hat for low brows,
        as serious as lipstick on the wet reed of a warm sax
                in the blue haze of a blues den.

All our instruments are lie detectors,
        detecting wounds that never heal.

I see myself

beginning in the compromise
             of their eyes, hugging them in spite
of atomic cocktails on their breath and plastique
             explosives between their teeth (they ask me
for a pick).

             The stifled anger bred in dank and dark places
                   under rocks and stairwells
presses hot against cool collars
             and emerges in a fist from the tuck and roll
                   of worried sleeves.

They nowhere go in moderation, hotwired
and piston-pulsed, hanging out with candied yams
             in the lust gardens of metromongers, lining up
             like stooges to take a fall,
             half-literate charity cases beating
             a limp legacy into blacktop.

We've taken the pisspots and sold them as wine
and they've taken the drink, turned it into acid
and poured it over our heads while we were sleeping.

It's enough to slow down a scrawny conjurer

too slow for suicide
             whose mumbo jumbo resurrects no dead
       and (unwilling to test drive new spells,
his hocus pocus low on fuel, too proud
             to call a doctor) scratches verses in the sand.

                   Still,
             no answer needs little English.

36

THE GROUNDED ASTRONAUT FORGOES HIS PENSION FOR THE GIFT OF PROPHECY BUT LEARNS TIME DOES NOT FLOW IN ONE DIRECTION

When the worried man locked in the clock
ticked off the tenth year of chiming in the works,
he faced a block of billy goats half-past quitting time.
Acting out psychodramas of compromise with their eyes,
a public parade of charity cases, lame of luck
with no one else to blame, looks into smoke for answers.

We settle for mediocrity, beating limp legacies
into vinyl floors, steel drums into altered states,
squeezing some profit out of wetdreams in cyberspace.
Between waking and dreaming, dreaming and waking,
there are black holes both near and far
to take us to some distant star.

To make some sense of hands, the worried man
turns them back, turns them back
all the way to bioplasmic fogs
stirred from the creation of new light
where we see ourselves
beginning,
where we see ourselves
dragging a stiff green belly
through the sand, some Genesis beast
skindiving in the silver-pleated sea
behind the fluttering lids of dreaming eyes
waking to a Larger Earth, living in the contact zone,
amphibians gulping air into new lungs,
breathing-in this new life
where we see ourselves beginning.

All the rumblings from the astral plane,
the urgent tongues of angels tied in knots

between tick and tock, ubiquitous calls
from angry Yahwehs ringing false alarms
in burnt-out cathedrals,
issued in earnest from forking branches of time
we presume
inevitable as what has been and what will be flows in a line
centered on a NOW that forever eludes us.
When the new age unfolds counting its converts,
we will not be able to distinguish
the ghost from the hologram
though we capture it in images to run and rerun
on a hundred different channels, breaking down the moves
and countermoves into a simple formula anyone can learn.
Soon the swirling eddies and backwaters carry us,
and we lie, blue in the face, on the bed
of a frozen lake, lamenting the loss of wonder.

Imagine being a virgin when the spirit-altered
warriors march into town like Sherman through Georgia
switching multiple selves in time with a drumbeat,
rattles shaking to simulate the sound of rain.
We live within the walls of Jericho and the first cracks
go unnoticed under the frenetic bugling of brass.
Where will you go? Who will protect you?
Stone temple pilots have hidden the maps
with the way out marked in red.
They worm away toting the last of the loot,
the last secret knowledge written on stone tablets,
down dark and silent passageways under the vinyl floor
haunted with the spirits of animals.

I too distort the vision – too angry at the anger,
a blindly minor god crumbling before
the purge of perfection.

There was a time, perhaps a dreamtime,
when the worried man between tick and tock

came awake in the works and shaped the earth
into a spaceship in a tacky blue bottle
and sold it to Ticky Tom, the peeper's son,
who worked overtime as a teller in the bloodbank.
When the war of worlds cleaned out all the cells,
and the time manager couldn't find his watch
before his wife found him with that witch
and layed a whip across his back
raising a little whelp, the teller of time
who lost the time, uncorked this new dimension
into which *ring around the rosies* we all fall down,
by which we all dance by the quick stream
where a hundred dooman stand on thin ice,
where the song grows sullen and still.

We must see the Self as we see the sun,
not directly, but as a reflection in a broken mirror,
shards of looking glass reflecting
shards of looking glass
beyond the tangle of saying no to what we mean.
We must take time to steal the tongue
as we steel our hearts, revoking all mercy
and learning to be tough to touch,
dreaming in resistance, swimming
into the current of hope against hope
measured as carefully as the tides, summed up
in someone's unified field theory.

We must see time as bifurcating again and again
in a dizzying spectrum
of possibilities and seize the line
before it mutters away
for no answer needs little English.
Our instruments are lie detectors
and the world makes liars of us,
liars of us all.

THE GROUNDED ASTRONAUT DANCES
WITH HIS DEAD FATHER BESIDE BLACK WATERS

Down
      the
           dreamhole
                darkly dancing

                                  I spin
in my wrung mind
      with that look in my eyes
         following you

dancing darkly

      where down goes up
      in a whirling
           wind
      I see you, a message
         in a broken bottle
      spilled onto the black sand
      emptying into black waters
      under a black sky with two moons

through the wormhole
      I spin into another world
      past the pelvic pacers
who swear we've been here before

hanging out with the dead
      collapsing in my own gravity
         looking for some live wire to touch
         some Gurvich effect of biophysical light
         to bring me back, a new man

If the past only illuminated the future, you say
we could ride easy on the sway back of NOW
knowing how to straddle the two extremes
    how to navigate
the narrow channel   between clashing rocks
(rocks more shadow   than substance, still enough
to crush the mystic meanders   of fatherless heroes)

Above us
    the giant chews on his jargon
and his word comes down in triplicate
that the world is wrong
                    more often than we are right
            In the twilight time
            while we're polishing the relics of dawn
good ole' boys out back, employing
poor and desperate Chinamen to tunnel
            under our house, keep digging up
the long-forgotten bones
we buried   After a lifetime on your knees, father,
                after a lifetime with blackface, eyes
                like a raccoon squinting in the light,
                breathing asbestos, smoke and soot,
                they took away your dreams with a needle
                and closed the doors to heaven
            and I'll never forget that look in your eyes
            pleading for me to help you die

            They uncover our most sacred places,
            forcing into light the lightless creatures
            from the humps and dumps of death mounds

                    between dreaming and waking
                    rubbing raw eyes

We wake
to the sound of picks and shovels

and grope in the dark
    blind in the dark
standing on the very brink of the black waters
    but they are too illiterate to decipher
            the message of your marrow

What words will keep us from the edge?

The right word at the right time
like the right tool for the right job

But all our tools are weapons
and the world makes liars of us

                Father, your pain so depresses me
        I cannot stammer a healing word
                sucking blood into my boots

                    Language, in me, grows arthritic,
        my song dispirited and chanting alone
beside the black waters where beginnings and endings
    get confused
            But what cross slung on the shoulders
of this hectic hero were not of his own making?

I free myself from guilt
shape my pain into a lightning rod
and send it down
                the dreamhole
commit it to the cool water down below
Waxing my wings,   I g l i d e out
                              over
                              homebound valleys
of ragged and rugged natives
tawny in loincloths   with an ear to the ground
                keen on the earth's internal rhythms
                listening at the navel of the world womb

42

for a stirring inside
where five pregnant women
wait by the open gate
dance in your open hand

I am not the man I used to be,
crumbling before the purge of perfection

My father's death humbled me
reopened old wounds like tears
in the fabric of space/time
many doors both near and far
still open

I enter blindly
   falling then rising,
   feeling the whirling wind
lift me up
   in a widening gyre   and I hear
              the quick stream run over rocks
              where I see myself beginning

              Waving my own fading colors
       as I soar above eccentric old women
                    scratching their baldheads

I map these many worlds
          without any wish to conquer,
              affirming some sense beyond the senses,
                    a place where I can still find him
                          near the black waters
                          sitting
                          in the black sand

I mark my own gradual decline with easy strides
and sinking my teeth
into the black loam of the Larger Earth,

aware the Larger Earth is aware of itself,
my troubled ascension becomes
mere counterpoint to d
                      e
                   s
                       c
                        e
                          n
                             ding
                                  notes of stars.

THE GROUNDED ASTRONAUT DOES BATTLE
WITH THE LORD OF SMOKE AND MIRRORS
WHILE THE CHILDREN BELIEVE WHAT THEY READ
AND BECOME WHAT THEY EAT

Soon the common man will have his day
      by the quick stream,
his crack of dream splintered
      from the blue beyond, coming back as song
riding over rocks no longer rocks
      where the obvious is no longer obvious.
The common man   with common sense
finding what we have in common, living on
common ground
grounded in the Larger Earth
      will have his day;
he will triumph over elevated godmen
with splotches of blueblood on their upper crust,
      ruling like PHARAOHS over a distracted milieu.

      We are small,
      so very small
      with too many on top
      sucking our blood into their boots

while a rage like gangrene moves up our spines
creeping up from down below,
caught in the middle, with nowhere else to go,
between rape and rape, sitting in the eye of the storm.

There are no more welcome wagons rolling down
candycane lane in this town. No more Mary Poppins
popping in and out of our bedrooms.
Disney's long dead and took the friendly fantasies
  with him;

Imagineers have taken his place
constructing deliberate realities out of dreams
sponsored by the prophets of profit,
where the greater good consumes its goods.

The nagging nogs have found their righteous trumpets
and soon they blare in our face from the front line
riding the thunder of wounded tom toms echoing
          in the streets,
a ghetto curse rising with bloodlust from angry alleys.
They nowhere go uninjured,
these rakes in rags lost on the road to Castalia,
blown out of a spitvalve onto cold sidewalks in November
by virtuosos of violence, playing scales of culpability
in a counterpoint of nature against nurture, predispositions
versus volitions, victim or victimizer,
who can say? – unable to place the blame.
A smiling carnival of carnivorous beasts,
ever creative but oh so antisocial, breaking out
of sewers and cellars, spilling into the evening news
with myopic vision and visceral might, new
Darwinian warlords
comparing their scars and marking off their territory.

And who will be responsible for what they imagine?

"Not I," says the sponsor, "not I."

When they speak, all they can say is
                    You made us this way.
This is how you wanted us. Beneath you, less than human,
                    stuck in the clockworks, nogs in the cogs,
the miserable manifestation of one more system
          breaking down.
                    Your greed made us strong.
Now fear us!
          For all you have taken can be taken back.

Auspicious merchants tending their wares
                are wary of them
            They tire of watching their profits gradually
disappear
                    off the shelves and under shirts,
                        hocked for another high.
They used to only take off their masks on Halloween
            so we could peep into the freakshow
                of their mixed up minds.
Now you see them everywhere; they nowhere go uninjured
invading the dreams of children, throwing themselves
in front of cars, baring their scars like service stripes.

I see this boy with a nose
bright as a toucan's beak
grimly smiling up at me as he eats
his lunch out of a body bag.
He says he was lucky to find some saint
with an incorruptible cadaver
and buries his face again in fragrant flesh.
I am not the man I used to be;
I turn away, applying facades like sunscreen,
a camouflage of sticks and bones,
denying wounds but bleeding in my soup.

Above us
            the giant chews on his jargon
and his word comes down in triplicate
that the world is spinning
                but no one is winning.

Gray on gray, Audio Animatronic minds,
making the innocent pay for their innocence,
oh, come on, you know them;
they are your children –
only mirrors after all –

smoke and mirror children who make compassion
look like a sucker's game,
children like drugged up steers in defoliated feedlots
packaged by the USDA and labeled as lean and mean
and good for you but filled with poisons
killing you slowly day by day.

       I too distort the vision,
           a blindly minor god
       tongued by hot tongs,
           coming down out of my Watchtower
       with backsliding monks, coming down hard
           thumping our Bibles and spreading
the Word
       that means more than man,
           that is more than sum of its parts.
Though our fever pops like firecrackers
in the startled mouths of whores
we pass through like drugged up steers
already drawn and quartered, wrapped in plastic
and fed to children for a dollar's worth.

Soon the common man will have his day
       by the quick stream,
his crack of dream
splintered from the blue beyond, coming back as song
riding over rocks no longer rocks
where the obvious is no longer obvious.

       The mind's polarity,
       worried over the world,
          shifts
       from the guideways of social determinism
       to the exigencies of free will,
       a flip/flop\
       objectified by foul weather
       in the kingdom of sleep,

a pendulum swing throwing
pontificators of progress
into a backward spin.

All our instruments are lie detectors
and the world makes liars of us.

I have walked on the moon
and the dust still hasn't settled.
From New Gnostics to New Warlords
from MAGLEV to MTV, caught in the middle
of a cosmic tidal FL-ux
echoing deep   in the bones
                in a place unfurling no flags.
The specter of change is enough to bring even the most stubborn
        leather-necked,
        thick-skulled
                warrior to his knees,
                        blubbering about his honor,
crumbling before the purge of perfection.

What song will keep us from falling?
The right song in the right time
like the right tool for the job,
but all our tools are weapons.

Just try to separate   the dribble   from
                the gibberish!

    Who will listen?
    Who will care?

"Not I," says the politician, "Not I."

                Weeper, get off your knees!
        Those bone-deep wounds are self-inflicted.
        You carry high your torch, wading knee-deep

into the mesh of humanity and call it total immersion.
See yourself see yourself as the author of your misery.
The human heart flutters like a mayday moth
in stainless steel claws of great manipulators
who profit from prophecies of helplessness.
An age of clear thought sleeps as pure potential
in a magical world that is not so magic,
but banished from consciousness, withdrawn
from our lives by the good intentions of saviors
who keep it hidden in caves, caves
guarded by a murk of minds no more than machine
for which the sum is merely equal to the parts.

The world at times implores me to deny any sanctity of sinner
and I tighten the noose around the necks of offenders and
blissfully invite observers to watch them swing from the dead
limb of a century-old cottonwood tree beside the once quick
stream.

The common man with common sense
finding what we have in common, living on
   common ground
   grounded in the Larger Earth
        will have his day one day.

Just wait and see.

THE GROUNDED ASTRONAUT, BETRAYED
BY HIS OWN WORDS, SEARCHES FOR GHOSTS
AMONG HOLOGRAMS BUT FINDS ONLY
THE JUGGLER OF LIVE BIRDS

Once

    a crippled little worm
    worked a sordid wonder
    on an orphaned boy's brain,
    speaking inside his head.

    DING DONG the worst was said
    and a rash word escaped in anger.

From

    the trill of the tongue he tossed out pieces of gold
    that changed into crude metal at the feet
    of mother mercy until totally spent
    the cache ran out too late to apologize,
    too late to take back
    the echoes which altered his sense,
    the teller picked up his tale and twisted it
    with a wit that had the torque of a tornado.

Still
  no one bled, though wounded
  as surely as the knight
  who fought the dragon
  for some small favor
  from the king's daughter.

    A meteoric fallout of half-digested words
        caught in the current and carried over rocks

dreaming along with stars in their eyes
but dynamite in their teeth.
Words that detonate,
words that blow down walls
and line up like mad elephants trumpeting,
back up traffic with late drivers
leaning on their horns.

How
can whole castles
come tumbling down
on so few words
so badly wrought?

The blindfolded knife thrower under the bigtop
is confident of his aim at the half-naked assistant
in sequins and frills spinning on the fortune wheel;
what was meant to just brush her ear in a rush
rips through her eye and into her brain.
The flaming aerialist turning somersaults in midair
puts on a show, a real spectacle, but no one
is up there
to catch him. By make or break until each crack
in the leather of their smiles is pondered into metaphor,
they nowhere go uninjured.
Do not hold these words against them;
the magician can squeeze some sense out of a wet sock
pulled out of a hat and released as swans
composed of light. We watch them flit and fly
without touching the ground until they vanish
in the vines, snagged on hidden hooks
and crammed into crowded pages by eccentric old women
lolly-gagging with tongues wagging,
unable to follow the intent
but wondering
why such small progress
after a lifetime on their knees.

Excuse me, that's not what I meant, exactly –
the meaning is not clear.

        The whale sputters through his blowhole
then sounds
        singing some underwater requiem for
a latecomer.
        I stand near the edge looking over, holding
a hot sax
        dripping saliva off the mouthpiece.
I've been paged
        from a session to document the crashcrater
        of another flyboy who took us at our word.

How
        could you lose the princess
        in the passion, in the rubble
        of an ill-timed purge of perfection?
        Bounding like a wildcat from a cage,
        all rip and tear,
        such simple sounds
        with such a hard edge.

Forgive the truth,
        it wasn't what I meant exactly;
        it wasn't what I meant at all.
        If you just let me put it another way,
        I'm sure the sting
        can be reduced
        to a dull ache.

    The world makes liars of us!

    Our language grows arthritic.

Yet,

     why worry over words
        as though every word is a world?

Do

     they change the seasons?
        Bring snow in July?

Do

     they hold the stars
     in their heavenly course?

Do

     they bend steel
     or keep watches ticking?
     extinguish fields of fire?

     For what it's worth,
        perhaps better a song.

Still,

     enough to live by.

THE GROUNDED ASTRONAUT RECORDS THE BIRTH
AND DEATH OF A FLYBOY
IN THE RAREFIED AIR OVER HOLLYWOOD

Up

from the first leap
of Spring
following mystic meanders
of luminous wormholes
springs   a flying boy
in embryo
with wings
                    wet and folded
                    bred in dank and dark places
                    by the quick stream.

The sight of him   drives
the   foraging   fillies   to trotting in the fields,
another slick hero   composed of light
                                          emerging from crawlspace
                  in the glistening glidepaths of slugs,
                          climbing out of the tangled tongues
of well-wishers whose best intentions
                  confined him   to the pit.

I saw him there
            as I sat by the quick stream,
meditating on the melting MOUNTAINS
            by the burned-out Cathedrals,
saw him climbing
            the fabricated walls of Tinsel Town
                  like a voracious vine,
wings   of silver wax   spun with gold.

Oh, come on now,
surely you must know him.

> He's a relative of a blindly minor god.
> With the WORD quivering on his lips
> > that could keep us all from the edge,
> > keep us all from falling, if he would only

let it go.
> But no answer requires little English
> and being politically correct
> > > does not offend.

"Where are you going?" I call
> as he rests on a ledge,
> > his wet wings drying in the sun.

I watch him climb   higher   and higher,
  onward and upward,   level by level,
  until his wings unfurl like flags
    declaring himself a separate nation.
They told him his wings were useless,
    designed for show, not for flight, of little more use
      than the false eyes of a peacock's plumage.

I am going to spread myself so thin on a thermal,
he says, *that you won't be able to see me.*
I will hang-glide over the glitter and glow
of the dream factory
writing my name on the wind.

> And so he does.

Oh, how he plays the currents to ride so high!
Expressionistic bold stroke of light on colliding bellies
  of cumulus clouds accumulating on the horizon,
    bright as a toucan's beak.

From that height, he sees eccentric old women
rocking in their rockers
                watching the waves pointlessly repeating,
                        waiting for the twilight time,
squeezing some sense out of a wet sock
                and flipping the Blacked-Out Pilot plug nickels
                        while the band plays on.
From the zeitgeist of his internal latitudes, he wrestles
        to a standstill the forces of aBsTRactIon.

And when the sun no longer glints off his wings
        he vanishes
                into night
                        just as stiff, working girls
                        hit the streets to exercise their hot habits,
                        bellowing at passing cars like crows in
eucalyptus –
        tantalizing women in tight dresses slit up the side
                dancing across the street, stopping traffic.

They say so much without meaning anything.

I think they should be politicians!

                You cannot see the flyboy now;
                        he vanished into thin air.
                Yet I think of him floating up there
                        unseen by those living only to retire,
                like the time manager who cannot find his watch.

Below him
veins of moving light gush from the heart of the city.

I imagine how beautiful we must seem

                                        from up there,
                how a little distance transforms

the fume-swilling consumers on their glory roads
into streams of tiny phosphorus dots
flitting their temporal art across a dark screen,
mayday moths fluttering against stainless steel.

Yet, when he's gone, he's gone.

THE GROUNDED ASTRONAUT, LIVING
IN THE ENDZONE, GRIEVES THE LOSS OF MYSTERY

Living in a zoo

                  is better than singing the blues
                  to work every morning.

       I was the man   on the moon.
       I brought home   moondust in my boots
           and left footprints,
                 size 10,
                     in a windless crater.
  They will greet the first colonists.

My job: to spoil your precious luna,
              slay your mystery goddess.

MISSION ACHIEVED!

       The moon is not made of cheese.
       Imagination is not reality!

           How far will we go in search of what?
           As far as Self can go outside of

Self

           before it sees itSelf in some

Other.

Living in a zoo, as I started to say,
with perverted monkeys beating off behind bars
would be as much fun as climbing Jacob's Ladder
with junior executives GOING   UP

singing the blues to work every morning
to get their   little   piece of   sky.

Above us
the giant drools black and white

                         spittle
down his double chin,
                the half-digested slogans
of a blindly minor god   posing in pin stripes
brushing off
some lower echelon suit trying to swing a desperate deal
  (as all deals are desperate to junior execs).

            I feel more at home in levis and sweat,
              now that these bloody boots have spoiled our
   luna.

I *am* living in a zoo with perverted monkeys
in monkey suits still needing to rise above
the working stiffs, those poor slobs like their fathers,
breaking down for poor pay on assembly lines
and feeling fortunate just to have a job.

The new American Dream is: Winning Lotteries!!!

You, dear friend, are the *guaranteed winner* of a Caribbean
cruise, we will provide YOU with your own CHAUFFEUR. His
name is BILL and he'll drive you in your brand new BENTLEY to
the luxurious SUNSET HILTON where you'll be staying in the
COMMODORE SUITE overlooking romantic white sands of
ocean beach.

            All you have to do is play!
            Just give us the number of your credit card

                  and you'll be on your way
                  to JAMAICA.

            BE AN INSTANT WINNER!

Imagine! First on your block to own a new body
or to invest in an alien mind modeled after a brain
in search of some *original* thought, some
NEW frontier.

Act now! Limited time only!
Why make body music with monkeys
when you can use your mind to lord it
over lowlifes.
You too can be a blindly minor god
in the new neuro-techno-info-cybernetic-synthetic
pantheon.

It's all a game you too can play!

Don't wait, or you'll be too late!

A swan wilting in cold light
beside the yellow stream where you see yourself beginning
approaches in total silence, its wet wings drying in the sun,
looking like it's been dipped in acid.
I hear America screaming.
"Not I," says the politician, "Not I."
Of thee I sing!

VIVA LAS VEGAS,
now that's my town!
The uncrowned pontiff signifier of the United States,
all glitz and glamour, show and glow, gleam and steam,
but no substance beyond the Strip,
with rich, green lawns blooming in the desert

but no water
for any fish.

So how do we get off the ladder?
How do we step down without falling?

You can't even pity the poor
when they play their own games
with such fervor, so full of envy,
plunging with gravity towards Absolute Zero
and self-consciously steaming head-on into dead-ends,
shouting BONSAI! BONSAI!!

Ambition murders compassion.

      It drives the quarterback into the endzone,
            breaks the earth down into yardlines,
     penetrates the secondary with a long bomb
          and sends you sprawling in a bombed-out
cathedral...

     It dresses up co-dependents as cheerleaders,
         carries its seed behind a straight arm
   with forms in triplicate to document its aim.
        It measures its progress with chains
  and goes on breaking new slaves.

Hear me, Builders of the future, building
a better tomorrow,
let bygones be bystanders with tickets exploding
in their hand, witnessing events beyond control,

             the wheel of fortune
             spinning out of orbit
             on the far side
of the BIG BANG

             where consciousness unfolds.

We walk behind
       a slick and stainless broker
       sinking his teeth in moondust,

    razing any memory of ever being animal.

  The cloak of civility ensuring our rise,
  we live worse than any animal in a den;

WE LIVE IN THE ENDZONE, behind a straight arm.
Another line to open with a block.
Another drive.
Another score.

Another loser
        summed up in someone's unified field theory,
        more machine than man, understood and repeated,
                the product of another assembly line
                stripping your gears on a down and out.

Between us and a better tomorrow
looms the barrier of separateness,
a Wall of Jericho keeping us in or out
with half-digested slogans and promises of prosperity

that singularize,

severing ties

naturally connecting us to willow and wind.

We are falling, perhaps, into *voodoo* magic
through layers of a sedimentary mind,
chewing our cud in cyberspace,
reciting rhymes in unison.

In the cities, the only news from God
is ciphered in the works of man,
distorting the miraculous
into symmetries of steel

        and God knows,
        I deplore symmetry!

THE GROUNDED ASTRONAUT ACCIDENTALLY RELEASES THE EARTH GODDESS FROM THE MEAT LOCKER WHERE SHE WAS FROZEN BY LOGICAL EMPIRICISTS TO TAME UNRULY PASSIONS

Gilded in gold,
> brewing a storm
> of answers for as yet unasked questions,
philosophers stoned on reason
> dare not be thrown by any twilight
> > so theorize how even accidents fit their model.

We let them foist their fragile tapestry
and examine the warp of their words, half-digested
> in courtly compromise of corporate vogue.

However the dance is done,
collapsing
around its own gravity, no man stomped
> a wiser tune into a goat's head.

I paid my dues to tunnel through the sedimentary mind,
> employing poor and desperate Chinamen
> to carry the picks and shovels.

> I walk on the moon, my footprints in moondust.

Proper paladins escorting lost apes into temple tombs
where salaried soothsayers interpret the mystic meander
of fatherless heroes sharpen stone knives to slice

through thick hemispheres and bring us closer to God.

We have evolved into a threatening species

living only to retire.

The quiet dignity of the wild
once captured
only by camera,
stealing the spirit of exotic beasts,
are now too common to behold

and stand spiritless.

What once was wonder now lives in a zoo,
kept behind bars and locked in cages,
pacing back and forth or round and round

but going nowhere.

The instinctual prowling of night-stalking cats
is reduced to circles on concrete.

As their numbers diminish, we cannot identify
their traits
when we uncover them in tangled forests of the self.

Behind the laminated civility   and neon advertisements
for city life with fast food addictions
with all its accessibility and convenience,

behind the hard-edged geometries

of glass

and steel,

our memory of the waterhole grows dim,

our memory of the cave grows even more remote
and non-threatening
– a cooly sublimated animal desire.

Before they closed the spaces and asked for permits,
before they directed and regulated and required
environmental impact statements,

> when the forest was full of trees and nymphs
> and the trees were all old growth with a mill girth
> sufficient enough to keep violent men working
> so they wouldn't abuse their women,

> no one monitored our acts against nature.

Now we conquer nature in our homes.

> What do these philosophers say to Little Briar Rose
> imitating sizzling Rita Hayworth redheads in slinky
> skirts for a night out on the hustle bustle town
> with some Marine mover perpetuating the war
> between the sexes by grunting over his steak?

> What do they say to the sad doll in her nightie
> weeping over whales while her uncle pounds her flesh
> with the same bloodlust that drives the harpooner
> to deck shouting *"THAR SHE BLOWS?"*

> She stiffens to his touch and arches
> her back surrendering to his deep thrust,
> held prisoner in her own body by

a perverted manhood

> spurting in fountains of fame
> and shaking the towers of Wall Street
> where Bankrupt brokers behind temples
> of glass sweep up their shattered dreams
> in the aftercare of another Crash,
> another petulant boy standing with cream

in his boots,
        another poor puppy who's lost his bone.

What they cannot measure must be myth,

            yet all the tools we possess are weapons.

These mad monotonous machines perpetuate their own myth

– the myth of control.

Unstaggered
    by an impossible leakage of laughter,
unstirred
    by the loud language of bellowing blues blasters,

    they are quick to bemoan   the quiet of the quelled
                whose passions are kept in a silver-plated chest
    with castoff hair-triggers and tiny tornadoes
blowing up bystanders   and torching
the temples of youth
            yet in their fabled frenzy when their reason fails
            they tear out rosebushes with their BARE HANDS!

How long can the arsonist
lick the flames at the devil's door
before his tongue finally blisters?

        When the sharp extemes are dulled into normalcy
                by the tyranny of Mr. Analytical Average,

the dark, feline grace of the goddess walks on tiptoe,
    light as a flute, for one false move that scratches
the surface
    attracts some predator who's gone too long
            without the smell of blood,

some stoned philosopher overcome and feeling suddenly
                weak in the knees.

The mild meditator constrained by his own law

moves

like fresh rain   over a drought-dry landscape
    soaking the troubled couriers of burn and blow.

Among the bristling tribes of firestarters holding
acetylene torches and exhaling nauseous fumes,

in the momentary lull between the furies, when all
that can be destroyed has been destroyed,
an uneasy peace prevails and if you seize it,
get beside it, close enough to deny the distance
between yourself
and its still and stoic center, you will recover
the irrevocable treasure of a mirror calm surface
reflecting the perfect possibility that sleeps in the deep
awaiting to be
                awakened by a whisper.

Too long I resisted the temptress
in favor of reading Whitehead in a white room.

Too long I've left the tea leaves unread
and ignored the angels dancing in the palm of my hand.

I waited too long for the mistress of mirrors
to turn her away now because her hair
is the wrong color, clipped short and wavy
instead of falling loose and long.

I dreamed it black and thick as a gypsy's

instead of thin and blonde and out of place.
But I've waited too long to turn her away.

The reasonable man confined to the circle of his reason
            must break out now or narrow down his vision
                    until his voice is a callous cover
for a fatal wound

            and though he's pleasant enough to children,

            he bleeds in his soup.

THE GROUNDED ASTRONAUT STRUGGLES
TO BE REBORN IN THE WIDOWS' WATCH
AFTER TOUCHING THE FACE OF DEATH

Skimming the blue lagoon with the mistress
of mercy before the cry of insomniacs
wakes you in your cold tower
where madness murders desire,
the coolest calm flies stricken by the hounds of heaven.
You sleep while the world awakes,
dreaming while the brittle break,
crumbling before the purge of perfection.

Beside the still running water, by the quick stream
where you see yourself beginning,
silently somber widows weep,
swollen with shadows, giving birth
to a swan in bloom bred in dark places.
They rise, composed of light, in sheer,
white dresses and walk the beach at night,
filtering moonlight through their fingers,
letting sea foam dissolve in their small,
so very small hands,
a substance they cannot grasp,
a meaning lost in the ebbtide,
rushing out in a panic,
riding over rocks no longer rocks
as the vagaries of vacuums fill.

When the last of the wild is willed to the lost
and the found is fond of walls that waver and fall
we utter a word that means more than man,
but too little too late 'cause the widows won't wait.
Gone are the gods of brimstone and fire
rising from fears and whirling up gales that spit and wail,

burning down temples and dancing round ruins
singing outloud

Can Johnny come out and play?

Another blacked-out pilot, baring scars like service stripes,
flies fast and furious with hyper-fibrillations
foreboding a dead finish at the sound of machines
breaking down.
You wrap yourself in a few drops of ecstacy,
hotwired and squinting at the sun
and plunge into the cold stasis of unremembering,
fresh stigmata from the tongues of angels
your only rite of passage.
The lost sea of Mars pollutes your beer and you stand
on the bar as though crossing a bridge in the fog.
The other shore was never more real; the closer you get
the closer you come to death. You see him waiting there,
declaring himself a separate nation, rubbing
raw eyes which peel back like black petals
of an apocalyptic rose exposing the promise of progress
as a crack of dream splintered from some refugee
living out of a bag
who longed to be a hero but can never go home.

You can almost touch his face,
forcing into light the lightless creatures
who flutter like mayday moths in the kingdom of sleep.
The hounds come running, sniffing your crotch
and wagging their tails.
Can Johnny come out and play?

How far dare we go in a game without rules,
making them up as we run down blind alleys
like trained mice in a magical maze?

It behooves the small urchin to hide

in some dark underground chamber,
but even a lightless womb
can dilate prematurely with rude awakening.
Somehow we attract the blinding touch
of angels flocking to the miraculous site
of one agonizing resurrection from the sands
of neutrality. The wise old man with fuzzy face
confronts the sagging crone in her midnight gown.
Her wild, grey hair flies off in all directions.

Her eyes are startled pheasant beat from the bush
by a pampered pup leaping in the groin,
exaggerating the scope of his genius
as speed jockeys thump the thin metal
of their dreams and crash into melting mountains.
But what he said to shrink her in his sight
you fail to overhear. Her siren wail
breaking through old static of too many secrets
marks our gradual rebirth
after a lifetime on our knees.
And in some sense beyond the senses
you remember the melody
but forget the words.

THE MISTRESS OF SPEED BECKONS THE GROUNDED ASTRONAUT AND BLACKED-OUT PILOT TO FLY WITH HER FROM THE HOSPITAL ROOM AFTER THEIR LAST BREAKDOWN

We battle-weary warriors, asleep in our cage
and counted among the converts, see you
behind that chameleon mask,

mistress of speed,

imploring us to keep our injuries fresh,
to show off our stigmata with a sense of pride.
We see the violence lurking behind the deadpan,
through the faddish fogs of fading purpose,
the rankled ranks of drugged and diluted minds stumbling
over cracked sidewalks lacking clarity to discern
inconsistencies in concrete under pressure.

    Among the lost and found live many

                          lobotomized misfits
      projecting their phantoms into shadows

                and falling asleep

        with a dull thud,   their pickled brains
in limbo
  from electroshock therapy and psychoactive drugs.

How many signatures of God have been forged
    in triplicate
    with the blood of angels   on admission forms
    so some blacked-out pilot
    can be zombied by institutional care
    and left neglected in some monkey ward with bars
    on the windows, left to scratch desperate messages

on the backs of playing cards to be slipped under the door?

We're like children
trembling in our beds with blankets
pulled over our heads   and the therapists tell us
we can believe in a future
without portent, so full of promise
selling buckets of rice to the ramshackle residents of hell.

We give up a little to save a little
so we can go on going on.

But when we look deep into the night,
we find you there,
sitting at the end of the bed.
And we cannot deny we are related.

Floating in awareness at the soft edge of sleep,
your eyes are searchlights
narrowed down to a point
penetrating the will of my walls.

We watch out for you as we watch for monsters
in the closet,
under the bed,
behind our closed eyes.

The tremolo in our voices
betray our fear that we might one day get caught
without arms, with no doors to lock.

In that moment, the world ends
and we see ourselves beginning.

We see ourselves beginning again,
soaring from blurred remote regions of mirage

where glass towers blaze in the sun
and the fathers of flight so vainly ignore
the futilely grounded, lost in lore.

We see ourselves coming back from the edge
with sentiments like sediments layered over time,
            hardened into emotions without motion
            or even rudiments of rhyme.

You're a gimpy warrior's game
            for those lacking courage
    to face themselves in the warped mirror,
            those unable to gather up splinters of heaven
reflecting melting mountains between twilight and dawn.

    They split the world into friend   and foe,
                    normal   or insane,
sinking their teeth into the soft earth and sucking her
            blood into octane.

I've seen them crash and burn,
blindly minor gods   flying fast and furious,
        attempting to break all barriers
                    with that look in their eyes,
        their convoluted contrails snaking
                across the sky in erratic patterns of flight,
                    onwards and upwards,
    no more than machine,
gutsy and erotic champions convinced of their authenticity

            by some bored beautician who drums the street madly
            and muffles a scream, half-concealing a low, guttural
                orgasm at death-defying stunts the daring do
                as they flutter like mayday moths
                in the endzone, skimming the blue lagoon,
                spinning down the dreamhole

where we see ourselves beginning.

Come now, you've seen them
taking off without a countdown.

The ones who sacrifice their limbs
for that look in their eyes,
vainglorious, high-steppers
so hopelessly in style mounting
immaculate machines groomed
for speed, defying the grave
intestinal test with cocky machismo
just to please you, dear mistress,
just to embrace you for a moment
for in that moment
is all
eternity.

THE GROUNDED ASTRONAUT CHIMES IN THE WORKS
AS THE MAINSPRING UNWINDS

When the mainspring   u n w i n d s
breaking the stasis
                    of cold unremembering,
refugees from a war of worlds
                    bred in dank and dark places
                                        form up in staggered lines.
Now and then old Grandfather clock   skips   a beat
(as all our tools are weapons and make liars of us)
                    so here and there   drops a day
                    between the ticktock of lives
looking back to the dawn of time
                    as through a macroscope at some macrocosm
where we see ourselves beginning in a far off canopy,
                    thumping our drums in a history of bones
                                        where language grows arthritic.

How long can words   collapsing
                                      around
                                      their
                                      own
                                      gravity
                                              string-together
enough sense in a paper storm
  to find its voice sleeping in stone
                    when the clock strikes dumb?

No longer **SOLID** as a rock
the sheer
  cliffs
  and deep canyons;
  the melting **MOUNTAINS** form layers   *of sleep*
  with dream on dream

in a time perhaps a dreamtime
warping the s p a c e of our convictions, defying gravity
and pulling us
                        into the juncture
                                    of mattereality
                                            where madness
murders ambition, where we m|e|a|s|u|r|e
our progress with chains
            and sing the blues to work every morning.
The mainspring wound too tight in test of tempers
finally gives and melts   matter into mind.

My voice is a stone tablet
                    [buried] in a b o m b e d out cathedral
waiting to be uncovered by some psychic archeologist
who can even hear
voices   echoing
            in the halls of hollow men,
their souls trapped in the wreckage of another break..k..down.

Sometimes I ride this point
                        of light intent on some Purpose,
{surrounded} by a catalog of mindless indulgence

illustrated
   in slick, full-color pages thrown up in a whirlwind
- to distract and detain   with brightly packaged offers
                    and guaranteed giveaways - if only

I join the *self*-immersion
            of a generation mesmerized by the f/r/a/g/m/e/n/t/s
                    without ever solving the puzzle,
                    without ever seeing a LARGER picture,
                        the LARGER EARTH.

Sometimes I seek this light
in all the wrong places, but enough gets through

the chinks in my armor to see myself beginning
by the quick stream,
by the rundown rivertown,
where our children worry over the world
and settle for mediocrity.

      Sometimes even the lame of luck
          steaming head-on into dead-ends,
             shouting BONSAI! BONSAI!!,
   are fortunate enough to have every crack
      in the leather of their smiles
  pondered into significance by stoned temple pilots
     flaunting their fresh stigmata like pedigrees
      and when the clock strikes dumb
   away they run
     not with a flash but a flicker.
Though they were only looking to add
       something new to their resumes,
they find a word that **means** more than man
and a man that is more
   than the sum of his parts.

   singing in the eye of the storm,
     counting converts and keeping score.

   I await that day when the springs are sprung,
   climbing out of the dungeon to higher ground.

Who will hold back the
    pendulum   swing
  when the mystic mad wind up marooned
      with scalps in their belts,
   beast blood on their breath?

   Erase my name from the mailing list.

    Don't call me; I'll call you.

I am small, so very small
       sitting on the runway at Cape Canaveral
       (as if it were my b o m b e d - o u t cathedral)
       catching cracks of dream, spl/inters of blue heaven
       still drifting down after the shattering sound
                    of Saturn rockets.

       I gather up the f-r-a-g-m-e-n-t-s to forge
           a new metal to **hammer** out a new coat of armor.
       This one will shimmer with a thousand reflections;
           it will be transparent and let the light in.
       I will be chameleon man, rainbow man.
           For me the obvious is no longer obvious.

       I say too much without meaning anything.
       I throw my voice in hymns,
           reciting rhymes but longing for unison,
embracing my own echoes as contact with other worlds,
           all a part but forever apart.
       "Out there is in here," I say.

       When the clock strikes dumb, the last man
           runs down, shell-shocked by loud billboards
           and warnings numb nerves

ignore.

THE GROUNDED ASTRONAUT, WEEPING IN HIS BEER, ENCOUNTERS THE MISTRESS OF MERCY BUT HAS FORGOTTEN HER NAME

The girl at the bar is always Mary
      and though men have used her   flesh for a flash
      and passed through her like ghosts,
         leaving only fresh stigmata as their signature,
      she endures belly-up in bed,
         pink with dawn, rising in a momentary blur.

Through this swan in bloom,
composed of light,
I see myself beginning.

   Together in my old Grand Am, the loose fender
   rhythmically shaking and rattling like some poet
        raging in reggae,
     bumps and potholes making poor steel sing,
I saw myself beginning in her eyes,
        born from the tides,
   rising in a momentary blur after a lifetime on my knees
   awaiting the twilight hours where the winter dark
   and hungry hawk dives down on helpless children
in bed
   swooning as the shadow swells.
            I see myself
   in her eyes on a wave-washed shore watching her rise
   doe-eyed from the tides, another born believing
      stones have voices,   another born believing

if you listen
closely you can hear   a lost language,
      the thoughts of wild animals leaping
from mind to mind

with a mad compulsion to chew off the chains
and escape their bondage.

In her room, nothing is the same.
Objects float away without gravity.
We writhe like fish out of water,
tangled in streams of light,
last breath spiralling upwards –
a cry rupturing in space.

Exploding somewhere beyond fractured mountains,
we rise and smoke in bed like freighters on fire.

The time manager cannot find his watch.

The earth is contained in a bottle.

The new world is a wall too high to climb.

Her body is a temple ruined by barbarians
.who destroy what they don't understand.
I do not want to be one of them
and yet I am.

But I'm not the man I used to be,
simply   rehearsing for a hearse
those desperate last rites   that led to last night.

She stands nude watching the waves through the window.
The moon shimmering through translucent necks of breakers,
seafoam like blonde hair blown back by the wind,

looking out, wishing she was faceless as the dead,
knowing when she walks down the street, the dogs
will come running again, tongues hanging out,
sniffing at her crotch again

and nothing will have changed.

Can you hear her flute, that sweet fipple flute?
It calls me into the forest marking my trail
in the mystic meander of another fatherless hero.
I hold high the torch to the call of the cannibal
and plunge ahead in blackness thick as wet wool.
Creepers dangle in firelight, throwing shadows
on the walls of the den where we lay,
tossing and turning in our sleep.
A beast entwined in light invades our dream.
Fancy flukes of humpback whales sploosh somewhere
in a tank and I can hear them crying out
– their last breath spiralling upwards.
The camel's back is broken from the weight
of too many ministers collapsing
around their own gravity
and the tearful tomes of quavering gents
and diminutive dandies raised as consummate artists
of consume and consume, forced to face a world
they did not make, and do not understand
why they should be held accountable.

Waving my fading colors, I too distort the vision.

I helped make them who they are, helped make them
        with cooperative silence, a real team player.

I so wanted to walk on the moon.
If only I could have taken you with me, Mary,
having been there, you would know what I mean

        how the larger earth is my home and how
        there are many holes in space, both near and far,
        without traveling to some distant star.

> The muted man still manages
> to slobber and the lout can fret for his freak -
> a perfect denture darling
> who sticks her eyes with pins.

> There is perversity in beauty, a cold, self-inflicted desire
> to maim the perfect body and bear some ugly scars
> which attest to hard-earned character.

But a world of sin is not so inevitable
though the girl at the bar is always Mary.
She may follow you home or threaten your wife.
She may give you hell or help you find
the invisible door to heaven.

For a fetish of words and phrases
I pound the sense out of stone.
The groundwaves crack my hermit skull.
Down the dreamhole I go spinning
when the last dance is done,
summed up in somebody's unified field theory
where masters go on breaking slaves.
He who thinks the world to end
lives dangerously between the full bodies
of thundermating Gargantuas who flick the moon
from glacier lakes with silver tongues.
I sit by the quick stream in the pink dawn
remembering the melody
but forgetting her name, was it Mary?

I often fear words will master me,
> ingesting syllables brightly explosive,
> disowning them if they blow up too many bridges.
Politicking small cliques for a good word,
> plugging holes in the night with acronyms,
> scratching verses in the sand.
I prop up the world with broad-shouldered nouns,

the bloodless beat of their Anglo-Saxon wings
soar in wide figure-eights above the desperate plain.

But the girl at the bar is always Mary

and I can never forget her name.   In her visitation
        of many forms,   temptations to let go
        before language grows arthritic in me.
        She comes pregnant with possible worlds
        (worlds through which dreams enter)
        and she takes me back to the melting mountains,
                takes me back to my home in the grass,
where the tools of creation are composed of light,

where I see myself beginning.

THE GROUNDED ASTRONAUT, THINKING OF HIS COMRADES, FOLLOWS THE WHITE RABBIT INTO THE DARK INTERIOR AND FINDS THE LOST BOY STILL BROODING AFTER ALL THESE YEARS

On the aftershocks of withdrawal we rode
the backs of white convertibles, our tongues
unwinding like ticker tape, gulping the fresh air
and dragging our still wet birth behind,
our video dream stamped and approved
by a dozen commissions with remote control.

     My esteemed brothers, our roads diverged
     and I find myself alone, hoping we can meet again,
     that we could all throw off the headphones, kick
     the silent addiction to the half-truths
     of advertising, unplug our hot-wired cradles and slip
     bongo-eyed into morning fog and salt air,
     riding our drum to where eccentric old women
     rocking in their rockers polish the relics of dawn.

The heel of my right foot is lame, my back and neck
stiff from too many days behind a computer. It hurts
to walk across the parking lot let alone
a five mile journey down through the rainforest
and up along the cliffedge to the ultimate view.
This stiffness more than means a body state.
So tuned to manifest our humanness;
all boundaries oppress. The shroud of sleep
still smothers me, the last illness holds my breath.

I'd rather walk in warm, soft sand on the beach,
wade knee-deep into spent waves or curl up in the shade
and drift in dream, than trust my vanity

with the ultimate view as all my instruments are weapons
and I would not want to murder that peace.

   But I see you there, white rabbit,
   urging me up in resistance to gravity,
   (the force of law and ruthless nature of middle-age)
   cajoling me down the trail where every world left
   behind is dampened and cooled in the dark interior.

The path descends.

 I fall behind
    shrinking in the shadows of the rain forest,

a salted slug
 diving into its own flesh,
   recoiling
   from thick immensities of mossy limbs.

Forgive me now, esteemed brothers, but I was simply curious.
How can we fly so fast and furious, leave so much behind?

I no longer feel sunlight on my palm.

 Can any brightness endure in this deep throat?

I listen for footsteps and bump softly into ghosts,
lost worlds resurrected in green silence.

 I follow the white rabbit,

      become the boy
      who hid
      his self-abuse behind a book,

  follow him down burrows

in the crotches of octopus trees.

A haunting sound reaches me from a distance.
The old themes erupt profusely,
clogging the slick guideways.

I somehow got stuck, must now get unstuck.
The old themes offer little more than static.

Sometimes the truth is simple friction.
The view is jungle rot and decay.

You are what you see.

I see my own breath in the middle of May –
formless, dissolving before my eyes –
and I have lost all sight of you, comrades.

Your aim flies true to its mark,
to the moon, to mars, to deep space;
you are never without destination,
running late, hurrying along, tick-tock, gotta make it on time,
never deviating or stopping to trace the shape of a leaf
or to search the wet runways for skittering Douglas Squirrels.

I dawdle hopelessly,   distracted,
                                          falling
                          constantly behind.

You reach tomorrow while I'm still today, waiting
        at the lookout for the lost boy to show up, wondering
        what's keeping him, thinking he's brooding again,
                ready to go back
                        by the time he arrives.
        I don't know how he gets where he is, poor refugee,

and where he is can't be called a place.
I only know there is a center to man
            that takes a lifetime
                    to recover.

They say you can always go back   the way you came
            but the signs   point in all directions
            and no true north
                    can guide you
                            out of the dark interior.

    The compasses are all   fixed.
    Mechanical failures all   around.
    The watches   broken.
                    But it doesn't matter really
                    because he cannot tell time.

I just keep following that low moan...

It has always been like this,
    going deeper
            and deeper
                    to avoid the light
                    or what the light relatively reveals.

            A brown house on Belmont
            in Grand Rapids, Michigan,
            with a monster in the basement
            some creature from the Black Lagoon.
            I left my lost boy in Never Never Land
            in a time of Little Holland tulips,
            left him stung by angy wasps, the air thick
            with the fragrance of apple blossoms.
            I remember in Oz there were apple trees
            that wrapped their arms around Dorothy

and wouldn't let her go back to Kansas.
Could we have known the brown house
on Belmont could raise such Cain?
Whistles blow in dim roller rinks,
toe-stops on waxed floors
sound like slaughtered pigs squealing.
A sapling ripped out by the roots
in the twisted light of foul weather
sent spinning over the rainbow.
Too late, too late, we learned, tick tock,
our living walls were only glass.

I'm thinking of you, father –
how you stole my childhood,
how I erased that part of my history,
blotted out your memory.
I close my eyes but cannot see your face,
only the monster in the basement
comes clear to me, a green, slimy cyclops.

I once lived in the center, but now I skirt the edge
looking back the way I came,
looking back on what was lost

like the invisible friend I grew out of
when judgement displaced imagination,
when I was told there were no such things
as monsters in the basement
and had to face the fact that monsters –
despite their obvious salivating
appetite for my pale flesh –
were something I created
and keep on creating,

though for me the obvious is no longer obvious.

Something I created hides
>   behind the closet door, behind shutters
>   in secret rooms of a house unexplored.

>   What you see is what you are.

I did not know why I was so afraid of becoming visible,
afraid of ticker tape parades, of riding
on backs of convertibles,
afraid of attracting too much attention,
until a man who spent most of his life in basements
took me down and helped me see through the dark
that it was you, father, lurking there.

The truth of an image is the image.
I rejected authority even in myself
for fear that I would misuse it like so many others.
Perception is judgement.
The monster in the basement was my fear of being
my father's son,
>   *blood is thicker than water.*

>   Forgive me, esteemed comrades,
but I was simply curious.
>   Why do we fly
so fast and furious?

>   The path is knotted with fists.
>   I fall behind,
>   afraid I might end my life in bed,
>   drowning in green mucous and phlegm,
>   coughing up blood
>   afraid of showing up and facing you,
>   afraid I might have to let go of the only emotions
>   that ever sustained me.

THE GROUNDED ASTRONAUT EXPLORES
MOVING BOUNDARIES ON THE OREGON COAST
AFTER A MECHANICAL FAILURE
CAUSED A RAPID DESCENT

So I give up angels (all such matter of wings),
let them drift   in

                                pure
                                perfection

                                        unanchored
in a harp strum,
           their unchained melodies lacking bass.

With so many shapes in the clouds
                          we surrender too much ground.
So I give up angels   (all such matter of wings).

  This is a day to raise the Dead,
  to go back the way you came,
  sitting on your drum.

  This is a day to raise the Dead,
  to descend again into the forest

on moss-ridden trails,
  as native as a fern,
where shadows suck sunlight
  through the waxy sheen of huckleberry and madrone,
into half-concealed voids cracked by slivers of sky,
deflected shafts driven through girth of hemlock and spruce.

I am a loose dog sniffing the air
nose in wet places, picking up a scent

long tongue tasting the wild, savoring
ᐟthe ripeness of the hour.

   I am a snake, fully touching the ground,
   at home in the rocks, finding fissures in the earth.

So I give up angels   (all such matter of wings).

Listen boys, this is a day to raise the Dead!

we outward go inward first
outside inside out
we twirl and twist
up and up and up
outward see with inward eyes
turn the outside inside
in a bellydance of twirl and twist
up and up
through mulch of dry stalks
Let us out,
we cry in the grasses' grip,
but nothing lets go without a fight.
We push
hindmost upward down
high and low searching for gods
head high while feeling low
we twirl and twist
up and up in spite
the down of gravity
with perfect nose
for even weight
electrified
translucent light

through bellies of leaves
going where only
light can go

So I give up angels
(all such matter of wings)

In the simple strum
                of afternoon
                stepping back from the anticipated edge,

with red-ripe eyes
and wind-streaked hair,
stepping back then
                        plunging
                                headlong
                                carried by gravity
                                down a hill
                                like Jack and Jill
                                lying on the wet ground.
I surrender to fingers of fern, touching and letting go.
                (Are they the tongues of beasts
                or the wings of angels?)

You do the hokey pokey and you turn yourself around.

                The tools of creation are composed of light.
                        Life itself is a moving boundary,
        a generative order of creative forms in free play.

Pulling tight the skin around my bones,
I dip my hands into a pool of cool images,

holding a liquid dream in the cup of my hands.
Slipping through fingers, what cannot be held
I learn to let go.
It is time to let go of pain and resentment;
they no longer serve me.
So much of life passes by nameless,

96

shadow worlds not chosen, still alive
but sleeping like Muchukunda in a cave.
How soundly he sleeps
untouched by the seasons,
the deafening crash of symbols,
the drumming of my bones,
the extra weight collapsing on my bed.

How well the lost boy remains lost
until he lets me find him.
I see the fear in his eyes,
fear of the monster in me.
I take the boy into my arms
and rock back on my heels (no longer hurting)
singing to the wind,
shaking a bone rattle
under a sky corrected in red
scaring off any challenge
to my bloodfeast as I rend him limb from limb.

      and though my feet go on i hesitate
      perfume lingers in aftermath of a warm rain
      musk like pine pitch pervades the mind
      a jay chatters in the low light
      the mother of mercy lies on a mossy stump
      vine maples insinuating vegetable thighs

      and though my feet go on i hesitate
      my winter breath grows wings
      flocks fly south with each exhalation

      a wind comes up
      i lean her way
      but turn to cold stone
      in a drumming of bones
      and though my feet go on i fall behind

the odor of sweet decay
the curtains pulled

we run
and run
and run
until our bellies pop

our thoughts   still live in flesh
each passing moment   drinks our blood

Will we grow tired of the taste of things,
the smell of things, the feel of things?
Will we no longer brood when we forget
the object of our brooding? Or did the object
of our brooding merely stand for something else?

            I must leave these places that shut out the sun
or go down in flames like the Flying Fortress that crashed
            not far from here, killing the whole crew.

At the trailhead, the Tillamook Work Camp boys are sweeping
the parking lot – lost boys, tough boys with the sleeves
of their blue shirts rolled up, leaning on shovels and rakes
smoking and cracking jokes but chilled at the thought
of descending into the dark interior for what they see
is what they are and the truth of an image is the image.

  Some keys strike dumb these mortal eardrums yet ring
  soundlessly some part of me that understands.
  I forget the ecstacy of some descents for fear of losing
  a fiction. I feel this fiction growing stiff at the end
  of my journey. Aware I'm already sweating these lines,
  aware of dreams compensating a lack of movement
  in my life, but unwilling to risk losing control

for a monster is trapped in the basement and wants out.
Thinking again I'm not the man I used to be;
that man is dead. He took off without a countdown,
breaking the sound barrier, and I know
he must have crashed somewhere.

The only evasion, esteemed brothers, is total immersion,
        but dreams still find you out,
        chase you down the dreamhole.
        My Mayday fallen hero finally came down to earth
        but must crawl out of the wreckage on his own

so we can give up angels (all such matter of wings)
        and find the way home.

THE GROUNDED ASTRONAUT DECLINES AN INVITATION FROM THE EX-PRESIDENT TO PARTICIPATE IN A MISSION TO MARS FOR HE DISCOVERS A SWAN OF LIGHT IN A CRYPT BENEATH THE RUINS OF A BOMBED OUT CATHEDRAL

Why the work ends at only the beginning
I'll never tell, but the ribbon of road
unravels and I string along,
following no certain truth
down the dark spiral, down the open mouth
of a cavernous conch through which dreams
enter, spilled into vast and intersecting planes
where each moment is a crossroads
and all the choices I have ever made
bring me to yet another
with no one else to blame,
and all the choices I ever made
have no power over me, brought to this
sense of wonder with the world,
each breath a bird,
carried by currents and eddies
to the eye of the storm where I am
awake in my dream,
watching the dark seeds I've sown reap
thickets of thorns
listening to music round and round
come out of angry horns at twilight time
surprising even me with its intensity,
bringing down walls of cathedrals.
And I had thought the passion gone!

Silly this seeming to wait
for enough belief to fail alone.

Still no use to cry in the crypt
for ladled love, poor prisoner,
such a sorry excuse for avoiding surrender.

Oh, to unanchor that white bird,
that swan in bloom composed of light,
to fly into the sun, so very high, escaping
the pull of gravity, free of the strange attraction
to solitary servants of a limited self.

A walk on water is a stand on belief
despite dust-collecting volumes of fact
in libraries built to house the papers of ex-presidents.
Whirling green gales of overlooked evidence,
flung like debris into corridors of great minions
decay the firm foundations of absolute truth,
drive the most confident and benevolent kings
to shrivel and die inside their concubines
before the earth moves, evidence come to life in my eyes
writhing with pleasure shocks
of genuine passion, melting down
the arrogant man, no more than machine,
until he no longer resists the sacred sword
that slices away the meat
as if it were soft butter
to let free that swan in bloom
composed of light.

It doesn't take a dunce dance to know
where the call to adventure is a stirring of wind
the way out is often a fine tuning within,
and dance as you may in a May-day dance
following the mystic meander of fatherless heroes
plunging with gravity toward the eventual ocean,

emptying into all that is,
always without finality,
I will find you waiting there by the quick stream
where the obvious is no longer obvious,
where I see myself beginning.

And should I ever fly again, I know now
I cannot fly without you.

David Memmott lives with his wife, Sue, in the Blue Mountains of eastern Oregon. They have one daughter, Liesle, who will always remind him of *The Sound of Music*. His roots have grown deep in the dark loam of the block-faulted, ancient lake bed called the Grande Ronde Valley, *Cop Copi*, the Inter-tribal Sacred Space of the Valley of Peace and Plenty, Land of the Cottonwoods. He is a graduate of Eastern Oregon State College where he studied writing with George Venn. He was honored with a Rhysling Award for Long Poem in 1990 and his first book of poetry, *House on Fire*, was published by Jazz Police Books. He is the editor/publisher of the Wordcraft Speculative Writers Series and was editor/publisher of the speculative magazine, *Ice River*.

of *HOUSE ON FIRE*

*In this collection the poet appears as a narrator of the fantastic in
the ordinary - of the supremely strange in the apparent normalcy of
life in our world...One cannot read these words without
apprehending them as genuine; nor can one understand the thoughts
and messages framed by those words as being other than personal
and unaffected.*

> \- Mark Rich
>
> *Magazine of Speculative Poetry*

*Employing both crystalline imagery and meticulous craft, <u>House on
Fire</u> showers us with incandescent poems. Cast up by a firestorm of
creativity, perhaps, but caught in that moment when the wind chases
them across the dark heavens and they fall, each an omen, each an
ember of personal mystery that flares before it dies.*

> \- Robert Frazier
>
> *Star*Line*

*Here we find no versified SF stories; none of Memmott's poems depict
generically typical SF situations. Instead, each poem weaves a
language-pattern correspondent to a soul in crisis: here, intersecting
and brilliantly colored planes of discourse slide past one another in a
(speculatively conceived) carnival of existential doubt.*

> \- Ignatz Mees
>
> *Science Fiction Eye*

Available from: Permeable Press, Wordcraft of Oregon and Chris
Drumm Books

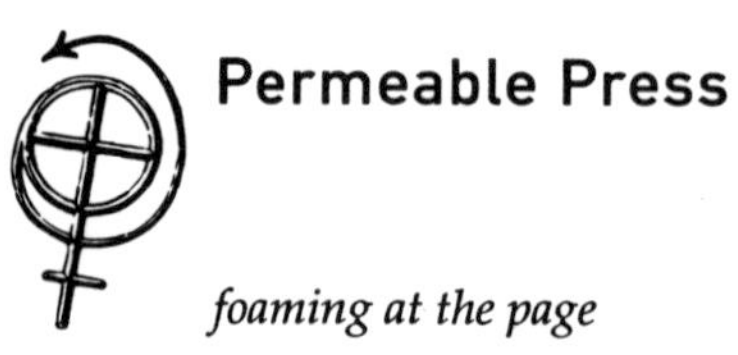

Permeable Press

foaming at the page